THE HEART
OF THIS GREEN BERET

memoirs and lessons from a life in uniform

Richard C. Towns

Author Photo by Dan Metts
Cover Photo by Lauren Clackum

Printed in the United States of America

Paperback ISBN 978-1-959563-11-2
eBook ISBN 978-1-959563-12-9

Maudlin Pond Press, LLC
P.O. Box 53 Tybee Island, GA 31328

www.MaudlinPond.com

ACKNOWLEDGMENTS

In Sunday School, I was helping Mrs. Jennifer Phillips in the class for 6th grade boys. They would come to class after 15 minutes of running between Church and Sunday school. So to calm them down I started class with some Army stories. I soon realized that I needed to end each story with a Christian moral to the story. I told these true stories from my experiences. I tried to keep them between 5 and 8 minutes, and I eventually ended up with 25 to 30 stories.

I had one young man, David Miles, who was sharp. He was writing down my stories and saying that he would put them together in a book. I have been thinking that if he can do it, so can I.

So I thank Mrs. Phillips for allowing me to be creative in her Sunday school class, and I thank David Miles for my inspiration for this book.

And finally, thanks to my wife Jackie who typed my manuscripts and edited my military terms so everyone would understand. She made it seem so clear.

Contents

HE BIT MY DOG SO I SHOT HIM

One day I got an urgent call from my wife. I could tell by her voice that something was wrong. An alligator had attacked our dog in the creek. We live on a saltwater creek on the southeast side of Savannah. Our house was built in 1895 and is on a dead end dirt road. It was a great place to raise a family. Up the creek from our house was the German Country Club. We were members, and during the summer everyone would gather on the dock to fish, crab, and swim.

Our dog was named Smutt, and he was my running companion for many years. In the morning darkness I would go outside, and he was always waiting for me. As we ran the sun would rise over the marsh. What a great way to start a day! Our children spent their summers at the German Country Club, and he would always walk with them from our house to the club. He would wait around all day, and then come back home when they did. There was a nice pool, and all the kids swam there all day long. On this particular day there were a lot of people on the dock at the club. Some were sunbathing and crabbing. I don't know if anyone was in the creek swimming but Smutt. He loved the water and always went in.

Smutt was swimming in the creek in front of the dock when a large gator grabbed and bit him. He started howling and trying to get free. Everyone started yelling and throwing things, hoping to distract or stop the gator. The gator then opened his jaws, released

Smutt, and started swimming away. At first, Smutt swam after him, then changed his mind and came back. Our dog was a 100 pound lab and king of the neighborhood. He had fought and won against all the dogs around us, and even an occasional raccoon or possum. I asked my wife, Jackie, if he was hurt, and she said just some puncture marks on his stomach. Everyone was scared as the gator swam away. They watched until he disappeared. Our children swam in this creek, and that put a stop to it! I thought that if the gator would attack a 100 pound dog, then he would surely attack a 60 pound child.

I started hunting this gator early every morning and late every afternoon. I finally saw him in the creek one day and shot him. He was nine feet long and weighed 150-200 lbs. I have a picture of my son, Doug, holding the jaws open and looking in his mouth. I later learned that this gator was going into people's yards and killing their dogs. Someone who was probably very technical said that I should have applied and gotten a permit. I was not waiting; **he bit my dog so I shot him.**

It seemed a waste not to do something with him, so I cut the tail into a bunch of steaks. The rest of the gator went back into the creek to feed the crabs. Our crabs were big that season. My wife did not want to cook the gator steaks and told me that she and the kids would not eat them. I finally persuaded her to cook two steaks. She thought it would take the wild taste away if she marinated them in vinegar, but the meat soaked up the vinegar like a sponge, and when she started frying them, the smell was strong. It smelled like the house was full of burned vinegar. I tasted some of the meat, and it was awful! I said, "Smutt will eat anything – give it to him." Smutt always ate our table scraps, and even went to the creek and ate shrimp that jumped on the bank as he splashed near the side. He even pulled up oyster shells and brought them to our backyard. With his powerful jaws he would crush the shells and eat the oysters. What a dog! We put the gator steaks in his dish, sat it on the sidewalk, and watched as he sniffed it and walked away.

Not one to give up, I called Harry Moraska, my Army buddy, and he came over from Goose Creek, S.C. Harry could cook anything for any number of people. He simply marinated the steaks in cooking sherry with layers of salt, pepper, and onions. He then grilled them, and they were great!

Smutt Swimming in Country Club Creek.

LEARN FROM A DOG

Smutt never went swimming in the creek again. He knew the danger came from the water. As you grow up there are places you should not go, and things you should not do. Your parents set boundaries by their authority, and you should not cross them. They love you and want the best for you. They also want you to learn from their mistakes. As you get older, other people will be in authority over you, such as teachers, employers, etc. Stay clean – do what's right. Don't engage in white lies, gossip, and slander. Turn away from sin – just think of ole Smutt and say, "I'm not going there, and I'm not doing that."

ACTS 14:15

Men, why are you doing these things? We also are men with the same nature as you and preach to you that you should turn from these useless things to the living God, who made the heaven, the earth, and the sea and all things that are in them.

I WANTED TO BE READY – HE MIGHT DROWN!

In order to be a Green Beret, there are many training phases and tests that you have to conquer and pass. Failure of any of these will eliminate your chances to earn the "Green Beret." One such test was the swim test. I took my swim test at Fort Bragg, North Carolina. We were about to enter "Phase Training" at Camp McCall, an old World War II Glider Training Base. We swam two at a time in full uniforms and combat boots. There were rescue swimmers who were ready to assist those who couldn't make it.

I didn't learn to swim until I was eleven years old when my family moved to Bona Bella which is south of Savannah, Georgia, and on the Herb River and Country Club Creek. I became a strong swimmer for short distances.

The Sergeant gave the signal to start as about 30 hopefuls looked on. I hit the water and did the Australian crawl until I needed my first breath. I continued on with several more breaths and realized that my legs were being weighed down by my boots. They had become sea anchors and were not helping me. I was swimming with my arms only, with my legs straight down under me. I also realized that if my arms gave out or quit, then I was going to drown! With determination I kept going until I made it. Even though I was totally exhausted, I looked for the guy who started out with me. I was told that he didn't make it. Those who made it went on to Camp McCall for Phase Training; the others went home.

Every year after that we had to take the annual swim test. We had to be ready in case we landed in a river or lake on our night parachute jumps. We later made ocean jumps which were quite an experience. We also learned to cross rivers with our gear by using our poncho's or pants inflated to float our gear, as we paddled and pushed.

We decided one year to have our annual swim test to coincide with our training in the swamps of Fort Stewart, Georgia, on the Ogeechee River. We had spent the day dragging an RB-15 (rubber boat for 15) through the swamps as we headed for the river. Boy, were we tired! We stopped at the river landing where you could fish or launch a boat. As the water came out of the swamp, it fed the river and curved around an island. The water was black but clean, and the current was swift.

So our A-team, about 12 soldiers, dove in and headed for the island. Our Captain stayed back to keep count and monitor things. That day he was going to qualify on paper, not in reality. As we were swimming, I noticed that Harley (we all called each other by last names because that was the name on the uniforms) was struggling, and I was very concerned. I kept an eye on him all the way to the island. Everyone was catching their breath and resting. After a minute, I headed back first. I got back and told the Captain that Harley might not be able to make it back. He was what we called a trainee. He was training with us until he went off to Phase school. So we really didn't know much about his swimming skills. I told the Captain that *I wanted to be ready – he might drown.* I took off my boots, socks, and uniform – right down to my boxer shorts – and I waited.

Harley kept waiting as all those around him swam back across the creek. He finally pushed off and started back. About half way back he started thrashing around in the water. He was slapping the water and going up and down. I immediately dove in and headed for him. As I approached him, his head was going down under the water. I went under and swam to his rear so he wouldn't grab at me and take both of us down. I got him around the back of his head and started talking to him, trying to reassure him and calm him down. He got still, and I started to tow him to safety. All of a sudden, he gave a big twist, slipped out of my grip, and went under. The water coming out of the swamp was dark, and I couldn't see him. I knew the current would sweep him away, plus the current in the stream might take him towards the roots on the bank. I felt this terrible fear come over me that he was going to die. I dove down and could not see more than one foot in front of me. I went quickly in the direction that I thought he was going, went down about eight to ten feet, and then went to my right which was the outside of the curve of the creek. I had enough air in my lungs for about 2 minutes and thought that this was my only chance to find him. Failure to get him would find his body going out to the main branch of the Ogeechee River and heading out towards the ocean. I'm not sure if I was praying because everything was happening so fast. I felt something, and it was Harley. He was drifting with the current. I quickly got hold of him around his neck and headed for the surface. With one arm around him, and one arm paddling, we headed with the current towards the shore. I remember touching the bottom in the shallow water at the bank. Some of my buddies grabbed Harley, and I laid down on the bank, exhausted.

I'm not sure how many of you have experienced an "adrenaline surge," but it takes all of your energy and temporarily zaps your strength. The Captain, who was sitting there, said, "Get dressed. Some people are starting to gather around." I was lying there in my boxer underwear and thinking that he does not know what just happened.

After that day, we never saw Harley again. He just disappeared. When I thought about it later, I realized that he never even thanked me – not that I needed it. We were trained to do what was necessary.

Not everyone has the opportunity to save someone's life, but we can all affect or change someone's life. As a Christian, you're different, and there are those who want to be like you. That's a big responsibility, especially when it's a young person. They are watching you, so don't disappoint. Your testimony can help others to lead a better life. Talking and sharing with those in need, or in doubt, can help them get through tough times in their lives. Finally, your prayers for those around you are a lifeline to heaven as you intercede to God on their behalf.

PSALM 16:1-7

God is our refuge and strength, a very present help in trouble. Therefore we will not fear though the earth gives away, though the mountains be moved into the heart of the sea, though its waters roar and foam, though the mountains tremble at its swelling. There is a river whose streams make glad the city of God, the holy habitation of the Most High. God is in the midst of her; she shall not be moved; God will help her when morning dawns. The nations rage, the kingdoms totter; He utters His voice, the earth melts. The Lord of hosts is with us; the God of Jacob is our fortress.

HE WAS BIG, UGLY, BLACK AS SIN, AND LYING THERE STARING AT ME WITH EVIL EYES

To become a Green Beret there are many phases and tests to complete. Fail one, and you're out. I attended phase training at Camp McCall in the far corner of Fort Bragg, North Carolina. It was a small glider training base left over from WWII. Very primitive, it had no running water, no electricity, and bunks of plywood. Of course – no mattresses. Classes were held in a tent containing wire cages with snakes that we might encounter while there.

I was the oldest in the group and was in great shape as a long distance runner. Having completed jump school and the swim test, I was ready to go. During this phase of training we were drilled day and night on land navigation. If you can't find your way around in the world then you're no good as a Special Forces soldier. We would be traveling on foot in all types of terrain, which included hills, forests, jungle-like conditions, and the good ole southern swamps.

My group was very diverse. We had men from California, New York City, Philadelphia, and my buddy and I from Savannah, Georgia. We settled in thinking the next day would be busy. Well, midnight came, and they woke us up for a fast paced march with ruck sacks. We got back, slept for one hour, and it was time to go again. They said to pack light because we were always moving fast. There was a high zip line that we climbed, held on with two hands, and away we flew to the middle of a lake. Dropping in the middle, we swam to shore, and this started our movement in the field. With wet uniforms and boots, we immediately headed for the woods, and we alternated leading to show we could read a compass. Moving cross country with only a compass day and night is not easy, but this was something we did back in our unit at Fort Benning, Georgia. Throughout the day we moved through the sand hills and then into the pine forests. We also worked on team movement, still alternating different ones in our group to lead us. All the while, we were being evaluated by the cadre (those in authority over us) of Camp McCall.

In each training cycle many soldiers did not make it. One day one of our guys got poison ivy on his hands and rubbed his face. His eyes closed up as his face got swollen. He was from a city in California and had not spent time in the woods. The instructor asked me to take him to a landing zone several miles away. I think he chose me because I was older, in great shape, and seemed to know my way around in the woods. We were already practicing the buddy system, so I chose Marco Vega, a tough little guy from New York City, to go with me. We had to have code names as call signs to bring in a medivac helicopter. I became *Georgia Boy,* Marco became *Spanish Fly,* and the sight impaired soldier became *Jungle rot.* Taking a bearing on my compass, we led him through the woods about two miles to a clearing where a helicopter was waiting. The medics loaded him and took off with dirt and leaves blowing all around us. We never saw him again!

As Marco and I traveled back, we passed a river and took a bath; I had a bar of soap, so we bathed. We were beginning to think like a Green Beret, as you never pass up an advantageous opportunity. As one evaluator in Germany told me many years later in Germany, *If you're not cheating – you're not trying.* We back tracked to where we left our group, got our bearings, and caught up with the others. When we got back, we teased everyone about how clean we smelled.

We worked for days setting up ambushes. Our days lasted about 15 hours, and we slept wherever we stopped – right on the ground with no sleeping bags or ground pads. Then we moved into the swamp. The big dangers were drop offs where you might suddenly go into a deep hole, and there was also quick sand. Land navigation becomes very difficult, and almost impossible. In a swamp you can never sight ahead to get a landmark for more than 30 – 40 yards. Also, you can't travel in a straight line, and you're always in water from your knees to your waist. If you know what you're doing, you can track the sun as it moves, that is – when you can see it.

For whatever reason, and it might have been my turn, I was chosen to lead through the swamp. On dry ground, my pace count is 113 strides to equal 100 meters, but it is impossible to keep track when you're trudging through the water. For example, when you approach an area you can't go through, and have to go around, you take a left turn for 50 meters, then a right turn for 100 meters until you pass the obstacle, and then 50 meters back to the right. Now you are back on the original azimuth, that is, if your pace count is correct, and your steps are consistent. You hold your compass in your left hand with the strap around your wrist. You continually sight forward to your next landmark as you hold it in front of your eye.

We came to a small stream running through the swamp. It was about 20 feet wide and 4 feet deep. I was leading, and the men were in single file behind me. Moving slowly, I was testing each foot fall as I moved into the stream. Water in swamps is normally dark because leaves, limbs, and stumps decompose. About halfway across, I looked forward and noticed the "wait a minute" vines. (Vines with thorns that grab and hold you) They were thick and had grown down to the sides of the two foot path.

There in the path I saw him. *He was big, ugly, black as sin, and he was lying there staring at me with evil eyes.* It was the biggest

Cotton Mouth Water Moccasin I had ever seen! In hindsight, I should have detoured and gone in another direction. I froze and signaled those behind me to stop. I looked at the snake, and he was staring at me with his tongue going in and out. I was thinking: We're in a bad spot. We're in water up to our waists, and this is where the snake lives. He's a powerful swimmer, and any minute he will slip into the water. We are the intruders and have invaded his world. I looked around and saw that my guys were in the water and backed up behind me. I knew that if the snake goes into the water, he might bite me – these moccasins are very aggressive. If I turned around to run, it might cause chaos as we stumble and fall over each other. I also remembered that I was being evaluated and should show some leadership. Over my shoulder, I called for a stick as I kept my eyes on the snake. As I was handed the stick, I noticed that it was covered with fungus and had been laying on the wet ground for some time. It was water-logged and heavy and about 8 to 10 feet long. As I was swinging it over my head it broke into several pieces that fell on my head and into the water around me. I was both mad and scared. The moccasin had raised it's head, and it's body seemed to be moving into a tight circle. He was still on the bank and hadn't moved towards the water or us. I raised my hand with the halt signal because some of the guys were moving around me so they could see. I tightened the strap holding my rifle across my back. We were only issued blanks, so it was only good as a heavy 3 foot club. Keeping my eye on the snake, I asked for another stick. This time it was a good one. As I swung the stick, I knew I had to break his back. I hit him hard, and he was convulsing and flopping around. I quickly raised the stick about 10 feet and came down hard again. The snake was twisting and moving, but not headed for the water or us. I quickly hit him 3 more times. Everything got quiet as I was standing there in the water with the stick raised in the air. I then laid the stick on the snake and applied downward pressure as I started walking towards it. Looking out for other snakes, I came out of the water and watched the snake closely. I determined that it was dead, so I picked it up with my stick and carried it to a clearing.

As a Southerner, I was amazed that some of the guys had only seen snakes in a zoo. I opened the snake's mouth and showed them how it was white inside, like cotton. The fangs are hinged and were dripping with venom as I pulled them down. If a snake that big bit you, and pumped in a huge amount of venom, a person might not survive if it took 2 hours to get to a hospital. I then thought about Walter, the other member of my unit that came with me to Camp McCall. He was infamous in our unit for a question he asked at a pre-jump briefing. Those jumping that day were huddled around the jump master as he talked about safety, the type of aircraft, jump altitude, and weather conditions. The jump master then asked if there were any questions. Out of the wild blue Walter asked, *Are the winds subject to change?* This question echoed through our unit for years as we, in a joking manner, would say it and chuckle.

Two days earlier Walter had caught a 4 foot copperhead snake and was carrying it in a burlap bag. This is a pretty dangerous snake, and he had it over his shoulder on the back side of his ruck sack. I knew he wanted to take it back to the head shack and put it in the cages where the snakes were kept. I pulled him aside and told him we were still out here for many days, and if someone bumped the bag, the snake would probably strike. If that happened, and someone got bit, then we were a long distance from a hospital. The antivenom shot would be needed as soon as possible, so I suggested he get rid of the snake. He then held the bottom of the bag and shook it. As soon as the copperhead hit the ground and began to coil and strike, Walter began chopping it with his machete. The snake ended up in 6 or 7 pieces, and we moved on. I said a quick prayer because God had protected us.

A week later we came out of the woods and walked into the main camp. We had successfully completed this course. We made it! The snakes didn't....

My best friend's mother was a Godly lady. She just had that presence about her. As you talked to her there was no doubt the she knew the Lord and walked daily with Him. She had a wood stove for cooking in the kitchen, and a wood stove for heating the house. She depended on her husband's salary alone because she was raising 7 children. In spite of having to be frugal, her kitchen was always open to anyone. As a regular in her home, I knew that she made the best coffee and coffee rolls in the world. She told me that when she felt threatened by the devil, she would say, *In the name of Jesus, I command you to get behind me Satan.* She got this right out of the Bible in Matthew 4:10, where Jesus tells Satan, *Away with you Satan! For it is written, you shall worship the Lord your God and Him only shall you serve.* When the evil of the devil tries to envelope you, just quote scripture and pray to God.

PSALM 121

I will lift up my eyes to the hills from whence comes by help. My help comes from the Lord who made heaven and earth. He will not allow your foot to be moved. He who keeps you will not slumber. Behold, He who keeps Israel shall neither slumber nor sleep. The Lord is your keeper; the Lord is your shade at your right hand. The sun shall not strike you by day, nor the moon by night. The Lord shall preserve you from all evil; He shall preserve your soul. The Lord shall preserve your going out and your coming in from this time forth, and even forevermore.

I PICKED OUT THE BIGGEST AND BADDEST ONE AND CALLED HIM OVER

Springtime came to Fort Benning, Georgia, and the weather was beautiful. We had finished a tough winter of cold weather training. We had slept in the snow at -15 degrees, and like good ole Southern boys, we were ready for sunshine, heat, and bugs. Our unit planned a family day with demonstrations of some of the things that Green Berets do. My family came over from Savannah on a Friday, and I booked them into the Days Inn. Little did I realize that jump school had just finished. Those young men had endured 3 hard weeks and were now *airborne*. It's one of the things in life that you always remember with pride.

That afternoon I took a break from the preparations for Saturday and Sunday and went to check on my family at the motel. My good friend, Sgt. Frank Rigelwood, came with me, and whenever he was around, you just handed him the vehicle keys. He was the closest that I knew to being a professional driver.

When we got to the Days Inn, Jackie, my wife, and the 4 children were not in their room. We then walked around the back to see if they were in the pool. We met them on the side of the building, and Jackie was not happy! I said, *What's wrong?* She said the drinking and language of the young soldiers was bad, and she didn't want our children exposed to it. I was livid, as I never tolerated bad language around my family. Frank was also angry, as he had known my family forever. As it turned out most of those soldiers were 18 – 21 years of age, and were showing off for my daughter, April. I best described April as 14 years old going on 19. Only 14, she looked older, was very attractive, and was even doing some modeling.

As Jackie stood there holding the hands of Erin and Zack who were only 5 and 3, I said, *Wait here a minute!* Frank and I walked around back to the pool and saw about 15 to 20 soldiers. It started to get quiet as they noticed us. I realized that they had never seen Green Berets. Well, the reputation preceded us as I walked to the edge of the pool. I was not happy and it showed. Frank, behind me, was standing there with his arms crossed and his jaw jutting out. He had tattoos on top of tattoos. He looked like he could kill! I was looking at each and every guy there. ***I picked out the biggest and baddest one and called him over.*** Actually, I yelled at him as I pointed my finger at his face. I said, *Get over here!* You could have heard a pin drop. He made his way to the edge of the pool and climbed out. He then stood at attention in front of me as I started a loud tongue lashing intended for all to hear. I discussed manners, discipline, and the pride of being a soldier. I reminded them that they were more than just soldiers. They were elite now that they were airborne. I ended by shaking my finger in his face with a warning: *If anyone here uses bad language or bad conduct, then you will answer to me!* He nodded his head, and I said: *Did you hear me?* He said, *Yes,* and I told him I'd be back in an hour.

I then brought my family back around to the pool and told Jackie that the soldiers behavior should be better. An hour or so later we stopped by the pool, and I asked Jackie, *How did it go?* She said the guy made everyone get out of the pool, line up, and apologize to her. He was not there anymore, and I guess he was not wanting to see me again.

SFC Richard C. Towns, Co B, 3rd Bn.

When I went off to college, I heard guys call their mothers *old lady*, and then trying to be cool and sophisticated, they also called their roommates *old lady*. I thought that one was disrespectful and the other dumb. These same guys, ten years later, were calling their wives *old lady*. Those guys couldn't seem to make up their minds. Looking back, I realized that my mother loved me and sacrificed for me to be in college. My roommate was my roommate, and that was that. My wife, that I chose to spend my life with, is younger than me, and I love her even more than ever.

Start now by honoring your mother, and later you'll treat your girlfriend and wife the same.

EXODUS 20:12

Honor your father and your mother, that your days may be long upon the land which the Lord your God is giving you.

GENESIS 12:3

I will bless those who bless you, and I will curse him who curses you; and in you all the families of the earth will be blessed.

IF I PULLED THE TRIGGER,
HARRY AND I WOULD DIE

Our commanding officer chose four of us. We had finished phase training, and we were airborne – we had earned our Green Berets. We were a close knit group, and we knew our stuff. Only a few were selected to attend this elite weapons school at Fort Bragg. Weapons from all over the world, including Third World nations, had been assembled. The purpose of this school was to acquaint us with weapons that we would come in contact with no matter where we were deployed.

From Savannah, we took my car because we received travel pay. I had a blue Thunderbird. It was big, and it loved gas. My buddy, Frank Rigelwood, loved to drive and was a very good driver, so I gave him my keys and said, *You drive.* Harry Moraska, Gary Wilcoxon, and I sat back and listened to the music. The car had an eight track tape player, and the quality and sound was good. I always preferred to sit back and relax with a cup of coffee and a newspaper.

Fort Bragg in North Carolina is a Special Forces Command Center. Our first several days were spent in the classroom, and nights were free. We were briefed on range safety and security, and then spent time disassembling and reassembling the complicated machine guns. We took apart the BAR (Browning automatic rifle).

It fires a 50-caliber shell which is large and devastating and has a long range. We learned to take apart weapons while laying the parts in a row. Then, when we put it back together, we just started with the last piece we had laid down and went in reverse with the parts. That way you don't end up with an extra piece, or worse, a missing piece to that weapon.

At night, we went into town for our meals. We made this a fun time after being so serious during the day. It was evident to others that we worked hard and played hard. A soldier from Ohio attached himself to us. That day after class we rode into town. We were listening to a country tape of Willie Nelson and Roy Price in the car. Paul Dutot, a soldier at the school, said, "What kind of crap are you listening to?" So I told Frank, *Anytime Paul rides with us, always play that tape.* We later heard that back in Ohio, Paul bought that tape, and his wife and friends thought he had gone crazy.

Our classroom instruction came to an end, and we were told to get ready for the firing range. As Green Berets, we always had our ruck and gear with us. We each had waterproof ponchos that could snap together to make a 2-man shelter, or one poncho could cover a man and his gear. They trucked us to the range, and said we'd be there for a week to 10 days. We were supplied with MRE's (Meals Ready to Eat) in vinyl envelopes. I never minded them because I could lose about 5 to 10 pounds a week when living off of them. By mixing the coffee, hot chocolate, cream and sugar, I made my version of *International Mocha.*

We set up camp on the edge of the range. Then a tractor trailer rolled up and unloaded a huge pile of ammunition which they covered with a tarp. It was guarded day and night by 2 soldiers. They had loaded shot guns and were not very friendly. Our instructors and the Range Safety officers then showed up with more weapons than we could count. These weapons dated back to World War I. They were from all the NATO countries and many Third World countries.

As soldiers that handled and carried weapons, we were in hog heaven. Few soldiers, and no civilians, ever got a chance like this. If we fired a weapon, we cleaned that weapon. Here's the good part: all of the ammo was free, courtesy of Uncle Sam, and ammunition is very expensive. Back home, some shells are a dollar a pop. Our days and nights were full and busy until midnight.

We would learn about a weapon, disassemble, inspect, reassemble, fire, and clean – one after another. The German machine guns from World War I and World War II were very well made and had killed thousands of allied soldiers. The United States also made some of the best weapons in the world. The .45 pistol hasn't changed in 100 years and is the preferred pistol with knockdown power. The M-16 rifle was an excellent combat rifle and it's still in demand today. The United States had a 30-caliber carbine like Audie Murphy carried in WWII. Our M-60 machine gun is still used today. We fired weapons with night scopes and some with silencers. Everyone wanted to fire the AK-47, and if fired on full automatic the front wooden hand grip starts to heat up and smoke. You could take it apart in the dark, as it only had 5 pieces. They brought us a small cannon from the Korean War. It was a 57mm recoil-less rifle. They had a stockpile of ammo that they wanted to get rid of. Bear in mind, these shells were over 50 years old, and we had a lot of misfires. When this happened, we yelled, "Cease fire" and everything stopped. The firing line was vacated, and only those at the gun were left. We put the gun on safety and opened the breach. We extracted the shell and carried it to a sump hole. If it exploded, it would be below ground, and no one would get hurt. Those were tense moments - hand carrying a shell that could explode at any time.

One night someone was stealing ammunition. It was stored on the ground, covered with a tarp, and guarded. They caught the thief by shooting his legs. They carried him off, and we never heard another word about him, except that he was not one of us. The moral of the story is: Don't mess with the Green Berets.

Days and nights flew by as we were very busy and totally absorbed in our activities. We remained highly excited as we put 'hands on' to all of these weapons. With our time on the range coming to an end, we finished with the mortars. These are truly weapons of wide spread destruction. One is basically dropping a rocket into a tube and firing it into the air. Sometimes you can see the target, and sometimes you can't. Your training with this weapon starts with an M2 plotting board. With this circular board you aim and direct the mortar to hit a target, sometimes with the help of a forward observer.

We spent a lot of time with the 81mm mortar. It was crew served, and we rotated as we learned each duty. We also fired it day and night. The hardest part of that weapon was the M2 aiming circle. It places you at the center of this board which is like a circular slide rule. You learn to plot your position in regard to the forward observer and the target which you might not visually see.

The infamous *School of the Americas* was held at Fort Benning every year in spite of the protesters. One of the officers from a mid-eastern country could never grasp the many concepts of this 81mm mortar and received a failing grade. Upon arriving home he was told that he had embarrassed his country and government and was executed. After that, no one failed. Anyone could be taught to set up the mortar and send rounds down range. It didn't matter if you hit anything – you passed.

We then moved on to the 60mm mortar. It was smaller and highly mobile. You could move it and set it up in a minute. There was a degree gauge on the tube, and you could tilt it back and forth as you're shooting at your target. As each round impacted, you could move the tube to get closer. When you thought it was aimed right, you pulled the trigger. Since each round had a killing radius of 35 yards, you simply rained down death and destruction with 3 or 4 rounds fired as fast as your crew could feed the tube, and as fast as you could pull the trigger.

My good buddy, Harry Moraska fired, and I fed the rounds

into the mortar for him. We then switched, and I started sending the rounds down the range. I always liked *one shot – one hit.* Anyone could fire multiple times and hit something. I wanted to hit the target and move on to the next one. I fired a few rounds and had my target zeroed in. You then fire for effect and send multiple rounds to bracket the target. When you have rounds falling all around the target, then everything there will be eliminated and destroyed. As I was firing, I realized that Harry had put a second mortar round on top of the one in my mortar. *If I pulled the trigger, Harry and I would be killed.* I quickly and loudly yelled, *Cease fire!* Everything came to a screeching halt. The range officer raced over to assess the situation. He cleared the entire firing range forward positions. That left me and Harry with 2 rounds in the mortar tube. Hopefully, they were not cooking off to explode at any minute. As we were taught at the beginning of this course, we now had to extract these rounds. The mortar projectiles look like a bomb or rocket with fins. They have cloth bags of black powder above the fins, and you tear off some to decrease the distance towards the target. When you pull the trigger, these bags explode and propel the projectile. We slowly and carefully started the process to clear the mortar tube. I cupped my hands around the top of the mortar tube. The fleshy part of my palms extended over the top edge. Harry then slowly and carefully raised the bottom of the mortar until the rounds started to slide forward. It was possible there would be an explosion that would blow apart the mortar shell and send shrapnel in a killing circle.

The top round stopped as it touched my hands. Harry froze as I extracted the first round. There is always the danger of a *cook off,* and it could explode at any second. I put it away from us in a safe place, so even if it exploded, we would be safe. Thoughts go through you head as you are staring death in the face. We then repeated the procedure and slid the bottom round out of the mortar tube.

What a relief! Death was knocking at the door, but we did not open it!

I believe if you belong to God, He will look after you. As a Christian, I believe in God. Does your life and actions reflect a walk with God? Have you made Sunday another Saturday? Where is your Bible - when did you last read it? How about your prayer life? Living a life in Christ can give you peace and comfort in times of trials. Seek God while he may be found.

PSALM 144:1-3

Praise be to the Lord my Rock, who trains my hands for war, my fingers for battle. He is my loving God and my Fortress, my stronghold and my deliverer, my shield, in who I take refuge, who subdues peoples under me. O Lord, what is man that You care for him, the son of man that You think of him?

HE BLACKED OUT A TOWN

We went into 'isolation,' a secluded and secure planning area at Fort Devens, Ma. Our A-team of 12 Green Berets were planning and rehearsing for an 'FTX', a Field Training Exercise further north. We divided the mission into phases that each member would research and plan according to his skills. We then cross-trained all phases so that the loss of a team member would not affect our actions. The day before the night we left, the top brass commanders came by and reviewed our mission, and we answered their questions. Upon their approval, we sprang into action. We destroyed any evidence that we had been there. After a final packing of gear, we headed to the airport after dark, carrying our ruck sacks, main parachutes, and reserve chutes.

We hot loaded from the deuce and a half (2½ ton truck with a canvas covered back) into an idling C- 130 transport plane. Within 5 minutes the plane was in the air. The plane was crowded, so 2 or 3 jump masters took soldiers back to the door area which was clear of gear. We started chuting everyone up. Bear in mind that it was night time, the plane was dark except for overhead red lights, and we were bouncing around. We safety checked all straps and catches of the parachutes, and then the men hooked up their ruck sacks (about 85 pounds) and their M-16 rifles. We sat them down according to their jump order. There were 4 teams on board, and they would jump one team at a time.

That night around midnight our team was hooked up and standing at the open door. The C-130 had slowed down to about 130 miles per hour, but we never got a green light at the door. Our pilot told the load master, a fellow crew member, that our jump was canceled. Our jump master, who had taken charge of the open door, got the same information. Our team of 12 was standing by the door and, because of the incredible noise, we were given a hand signal of a hand across the throat that our jump was a *no-go*. We unhooked our parachutes from the static line and sat down.

There were other jumps earlier from other planes and evidently an unsafe condition existed on the drop zone.

On the airfield after landing, we immediately loaded into trucks to be moved to a drop off point that was close to where we would have jumped. We had to move fast because daylight was coming, and we had miles to travel. Day break came, and we took a quick break

That's when we heard that one of the jumpers from the other plane had drifted to the edge of the drop zone near the road. They don't know if the wind caught and carried him, or if he was one of the last to exit the plane. Sometimes the last jumper would go past the drop zone. Under normal circumstances, you could encounter trees, or sometimes a road or highway. If there was water or power lines, then you could be in trouble. They told us that one of the guys from the earlier plane had crashed through a high power line, and **he blacked out a town.** That's all we knew, so everyone assumed that he was dead.

We completed our mission, and eight days later we headed back to Fort Devens. It was then that we learned the whole story: The jumper had lowered his ruck sack by pulling a quick release. This drops it about 15 feet below you on a tether and allows it to hit the ground first. This way your legs don't have to bear the brunt of an extra 85 pounds, and the jumper is less likely to get leg injuries. As he came down, his ruck sack went on one side of the power line, and he went on the other. His body weight and that of his ruck snapped the 12,000 volt power line. He was lucky to be alive.

There are so many scenarios and possibilities of danger when jumping, and we try to be ready for the unexpected. Before every jump we gather in a group and go over what to do for various landings. With a tree-landing in a forest, we don't lower our rucks, and we tuck our elbows tight against our sides with our closed fists in front of our faces.

When there is the possibility of landing in water, we drop our steel pots (helmets) and unhook the straps of the parachutes while holding the chest strap with one hand to release when hitting in water. When headed towards power lines you don't release the ruck. You put your feet together, with hands and arms inside the risers (the main straps from your shoulders and chute), and hope to clear the line or at least bounce off.

Why did he release his ruck sack? If there was no moon, it would have been dark, and he possibly wouldn't have seen the power lines. Well, it almost killed him! (We train over and over so that our actions are reflex. A jumper may only have seconds in some circumstances to make a decision that will save his life.) And we never knew...

Richard Towns, Dana Edenfield, Don Mew.

THE BEST LAID PLANS OF MICE & MEN....

We plan things and do things, and sometimes they just don't turn out right. We pray for things and our prayers aren't answered. It's said that God doesn't give you what you want, but what you need. You need to know our God is in charge – the Bible tells me so.

ISAIAH 55:8,9

For My thoughts are not your thoughts, nor are your ways My ways, says the Lord. For as the heavens are higher than the earth, so are My ways higher than your ways, and My thoughts than your thoughts.

July '82

Parachutist Cuts Power

JEFFERSON, N.H. (AP) — About 200 people were without power early Friday after an Army Special Forces reservist making a night parachute jump brought down a 12,000-volt power line.

"He's a lucky fellow. He could have been singed if he had gotten tangled up," a utility official said of the soldier, who was not hurt. The official also said the line had enough power to kill the reservist.

"He apologized," said John Pearson, district electrical superintendent for Public Service Company of New Hampshire. "I told him not to apologize, just to thank his lucky stars that he came out all right."

Pearson said the soldier had an equipment pack linked to him by a 15-foot tether, and pack and soldier came down on opposite sides of the line. The incident, which happened about 11:15 Thursday night, knocked out power in part of Jefferson until 12:30 a.m. Friday, Pearson said.

Ocean Jump

IT WAS NIGHT TIME,
AND I WAS IN A SWAMP

As a Green Beret, most of our jumps were in the night. That way the enemy can't see you coming.

So many things can happen, such as water, woods, and power lines. We had trained until all our actions and reactions were automatic. When we jumped with combat equipment, we carried everything that we needed for a mission. We were wearing our web gear and attached were a canteen, 2 ammo pouches, a compass, a combat knife, and our M-16 rifle. I also carried a Camillus 1980 multipurpose knife on my pants belt. My gear ruck sack contained everything needed for a mission of about 7 – 10 days. I always traveled light so that I could move faster, so my ruck weighed only 75 lbs. I knew what was in each pocket and could find it with my eyes closed. With everything – my parachute, the reserve parachute, steel pot helmet, ruck sack, rifle, web gear, and food – it equaled my body weight. At least when we hit the ground, we left our parachutes behind. That was about 40 lbs. And we moved out because you never lingered at the drop zone – someone might be coming.

On a training mission one night, my A-team flew to Fort Rucker, Alabama. It was a big drop zone, and we had plenty of time to exit the plane. Coming down I knew I was in trouble, as I saw trees and water below me. When anticipating landing in trees, you never lower your ruck sack. As I headed toward the trees, I kept the ruck hooked and hanging in front of me. That way a limb can't go between your legs and possibly break bones. Sometimes your chute can catch on a tree, and there you are – dangling in the air. That is a problem in itself! That night I crashed through the trees and landed in the water with only bumps and bruises. The roar of the plane was fading away, and there I was. ***It was night time, and I was in a swamp.*** It got quiet, and I heard none of my team members. I hit the quick release on my parachute harness, and it came off, including the reserve chute.

I carried a folded kit bag under my leg straps, so I stuffed the main chute into it. I then clipped the reserve chute to each handle of the kit bag; I found a tree and hung the chutes there. I had no way of knowing how far I was in the swamp. I was thinking everything out carefully and decided to leave the chutes, take my ruck sack, and get out of the swamp. I would then try to come back and get them later. There was no way I could carry everything out at night. I was in water – in a swamp. The ground is unpredictable, with bogs, and sometimes quick sand. My gear and weapon had to go with me – they can keep me alive. I popped a chem light and hung it on the parachutes. I then got my compass out and plotted an azimuth reading that would hopefully take me out of the swamp. The military compass is incredible – it glows in the dark, forever. Then I struck out and made it out of the swamp.

Bear in mind: it was still quiet, and no one was coming to get me! I put my ruck and steel pot on solid ground, put a chem light on top of them, and got a swallow of water. I then put a reverse azimuth reading of 180 degrees on my compass and headed back into the swamp. My rifle and web gear went with me. My rifle was always with me. I slept with it, and it was never out of my sight. Even though I was following my compass, I couldn't walk a straight line. I had to detour around trees and places that are impossible to walk through. However, I knew the chem light on the parachute would be a beacon in the dark. I was only about 10 feet off of my mark when I saw it. I thought, *I just saved Uncle Sam a couple hundred dollars!*

My third trip through the swamp was a charm. I got all my gear and headed back to the drop zone. Further down on the edge of the swamp, I found my men. They were gathered around a single tall tree. There, about 50 – 60 feet up was hanging our new medic. His chute was caught on a limb, and every time he moved it felt like he was going to fall. When it's pitch black, and you're hanging in a tree, you're supposed to drop your helmet and count the seconds until you hear it land. It could be embarrassing to discover you're only 5 or 6 feet off the ground. The next thing you do is swing back and forth until you can grab the tree trunk. Then you unsnap from the parachute and slide down the tree. If that doesn't work, you pop open your reserve chute and climb down the outside of the parachute. Don't climb down the inside, or you'll be in a giant bag, and you'll have to cut your way out.

So, there was "Doc" Smith, swinging in the breeze. He was a most important member of our team. Not only was he a Vietnam veteran, but he was also a *Goatlab medic*. Don't ask me what that is – I don't want to document the details. Well, Doc got down from the tree, and I got out of the swamp. Another good day's work, or should I say, good night's work?

A Spot of Tea during Ruck March to RON Site -South Wales.

PRAYER

I always prayed as soon as I hit the ground – unless a call to nature was pressing after being shut up for several hours on a plane. Talking to God is comfortable even when praising Him, thanking Him, and not always asking for something. I've often thought that God supplies what we need, and not what we want. Try it – even if it's just what I call a *bullet prayer* - like I taught the boys in my Sunday School Class where everyone would say a one sentence prayer and shoot it up to God.

PHILIPPIANS 4:6

Do not be anxious about anything, but in everything by prayer and supplication with thanksgiving let your requests be made known to God.

Night Jump in England.

I NEED 12 IDENTICAL BOOKS

We were heading to Fort Sill, Oklahoma. The Army, in all it's wisdom, was going to try something new and different: they wanted to blend in with the Green Berets large troop movements. We were trained to jump in at night, travel fast, hit a target hard, and be gone! So it was not a good idea! My A-team of 12 was highly trained, and each of us was capable of leading the team to complete our missions. We were cross-trained and knew every other team member's job. If something happened to one of us, we would divide his essential gear, call for a medi-vac, and carry on. This happened one night in Germany. Our demo sergeant got wet during a cold storm and got hypothermia. We called for a medi-vac chopper from Skullthorpe, England, divided his explosives, and pressed on with our mission.

After arriving at Fort Sill, we were assigned a barracks and went into isolation to study and rehearse our FTX (Field Training Exercise). Our team leader was a *hard charger*. He was a cowboy from Texas. He liked me because, even though I was older, he could not outrun me. I had also started a team fund in order to buy things the Army didn't provide. This fund would accumulate, and we were able to get some high speed civilian stuff. As Green Berets we would sometimes do things, and get things, and just not talk about it! While in isolation we were going over some old spy tactics. In fact, so old that some went back to the Civil War.

Since I controlled the money in our team fund, the captain told me, **I need 12 identical books.** I was thinking: *Where can I get 12 identical books?* Even if I went to a book store, they wouldn't have 12 copies. To make things harder, the books should fit in our ruck sacks. In my ruck sack, every pocket was filled, and I could locate every item in the dark. That's 75 lbs of essential gear for survival. As I walked across the compound, I saw the chapel and thought I would go there and pray about it. I was thinking that I need God's help – He will provide.

I went into the chapel to shoot up one of my bullet prayers – short, to the point, and straight to Heaven. As I looked around, I saw a table with Gideon New Testaments. They were black, and I knew they were for soldiers. The Gideons put them there as giveaways. The size was perfect for our top jacket pockets or ruck sacks. As a young boy, I had seen a Bible that a soldier carried in WW II. He had carried it in his top jacket pocket. It had stopped a bullet that was heading towards his heart. It was a New Testament, and the bullet was still in it. The verse where the bullet stopped had a big impact on the soldier's life. Well, I gathered up 12 New Testaments/Psalms & Proverbs, said a quick prayer, and headed back with my treasure.

Back in our team room, we handed out the Bibles to all 12 team members. We then practiced the spy technique for sending secret messages. It requires the sender and receiver to have identical books. You'll end up with a series of numbers that a courier will carry. For example: Luke is the 3rd book, so it's number is 3. Fourth chapter is number 4. Verse 7 is number 7, and the 7th word is "All." Therefore, to get the word, all you need is a military New Testament and the numbers 3, 4, 7, 7. The secret message will end up being a page full of numbers. You then send this by a courier who does not know what book was used. When we finished with this exercise, we all put these pocket sized New Testaments in our ruck sacks.

It gave me a warm feeling to see God's Word being opened and read. A short time later, I saw 2 of my men by the campfire one night, reading and talking about the Bible. I was thinking that you cannot read God's Word and not be affected. *Isaiah 55:11* says, *So shall My word be that goes forth from my mouth; it shall not return to me void, but it shall accomplish what I please, and it shall prosper in the thing for which I sent it.*

Traveling in a C-130.

My advice is to get a Bible, a real Bible – not a tablet or other electronic device, but one you can hold and turn the pages. Don't have one? Go to the bookstore and buy one. Yes, they have Bibles; it's the number one best seller in the world. And get a chart to help you read through it in one year. It is guaranteed to change your life and your world.

LUKE 11:9

"So I say to you, ask and it will be given to you; seek and you will find; knock, and it will be opened to you."

THERE WERE THREE BUZZARDS, AND NOW THERE ARE TWO

One hot summer day my Army unit was training at Fort Benning, Georgia. We were scheduled for an airborne jump, and a C-141 airplane was flown in from the airbase in Charleston, S.C. This was the only jet engine plane that we jumped from. It was a cargo plane and could slow down to 150 knots for us to jump from. Our unit had about 8 to 10 Jump masters, and we rotated being in charge of the jumps.

With this plane, you could jump from either side door or at the ramp at the rear. A ramp jump was kind of neat since it was a *walk up and step off*. However, it was harder for a Jump master, so we usually chose the side door. Of all the Jump masters, I was chosen that day. This would be a "Hollywood" jump -something we didn't get very often. We would be jumping during the day with no equipment or combat gear – a rare jump for us. We drew our parachutes and emergency chutes. After checking the date in the small log book to determine when they were packed, we started to put them on. After we were chuted up, the Jump masters checked everything on the chutes, and how we were wearing them for our safety.

We gathered each A-Team in circles and practiced the "What if" drills. We went through all the scenarios we might encounter in the air. What if we have a mid-air collision with another jumper? How about landing in water or forest? What do you do if you're heading towards power lines? Finally, what to do when you land on the drop zone. We went over the five points of contact as we would hit the ground.

Looking around, *Everyone ready? Let's load the plane.* We walked up the ramp and sat on the outboard seats (next to the side of the plane). The plane's load master is now in charge. We were snugly sitting next to each other, and he checked that we were wearing our seat belts. I continually walked

around, checking on my men, always looking at their chutes and straps. I wanted everything right with no twists in the shoulder straps. One of my guys was rigged wrong, and his shoulder strap on one side was under his arm. I stood him up and corrected it, and I told him, *You can thank me for an unbroken clavicle.*

Everything looked good, and we were off. I then told the load master, *Keep it cold back here, and no one will throw up.* He thanked me because, once we jump, he has to clean up everything in the back of plane. We reached an altitude of 1250 feet above ground level. The plane made a wide circle, and I got a 20 minute warning, which I passed on to my men. The pilot now slowed the plane down to 150 knots (about as slow as this big jet can go.) We always get a 10-minute, then a 6 minute warning.

At this point, the door was opened by the load master, and he turned it over to me. I quickly inspected the door and ran my fingers on the outside edge. This is where the static lines will be pulled as the men jump. A frayed or rough edge could cut the static line which pulls open the parachutes. With a quick look out the door, I checked to see that no plane was following us. You don't want men jumping out and colliding with a plane behind us. Yelling and using hand signals, I stood the men up and told them to hook up. We then went through the commands: check static lines, check equipment and sound off equipment check. The last man in line slaps his buddy in front of him on the rear and yells *OK!* This was passed up the line to me, and the first jumper yelled *OK!* and gave me a thumbs up. I then yelled *Stand by!* and pointed beside the door. I then went back to the open door and checked behind us and looked forward to the drop zone. The load master leaned forward, tapped me, and yelled *One minute!* Turning to my lead jumper, I yelled, *Get in the door!* as I pointed there. He stood there with both hands on the outside of the door opening, and I got close and tight to him with my left hand around his waist.

Now the C-141 is the only plane you don't jump out vigorously. It slows down to 150 knots, and you step out. The perforated steel door is breaking the wind, and you are swept along beside the fuselage and out to the wild blue yonder. If you jump straight out past the door, the wind will grab you, and the straps over your shoulder, around your waist, and between your legs will give you a king-size squeeze that you will not forget. Then your riser lines will be twisted, and you waste precious time twisting around to unwind.

Meanwhile up front, the plane was lined up to go over the drop zone. The pilot was looking for a ground signal that it was okay to drop us. He had slowed down from 300 knots to as slow as he could go. He then said to his co-pilot, *There were three buzzards, now there are two.* He had felt an explosion on his wing and felt a shudder in the plane. He quickly checked his engines and instruments. His flaps on one wing were not reacting correctly, and he told the load master to get everyone with a parachute off the plane ASAP! The load master told me, *We have a problem – put everyone out including yourself.* Sometimes the jump master is just along for the ride and doesn't jump. But I always jumped after my men were out the door.

We got to the leading edge of the drop zone, and I spotted the *L* on the ground (made up of panels). Quickly, I looked at my men and saw that they were ready, so I signaled Okay and slapped the jumper in the door on his rear and yelled *Go!* The men knew to follow him out at one second intervals, but we're normally quicker than that. I gave the thumbs up to the load master and stepped out. My body was slung forward, and then the chute inflated over me as I swung back under the canopy. I quickly checked the canopy for any blown panels and was assured that it was inflated. Then I checked for fellow jumpers to avoid a midair collision that could kill both of us. Now for five seconds I was floating through God's air and steering my chute. The closer you get to the reception party on the ground, the less you have to walk. With a good landing, I usually do two things: pray, and use the bathroom (not always in that order). After landing, I unhooked my chute quickly. Sometimes the wind re-inflates it and can drag you all over the drop zone until you hit the trees. I then S-folded my chute and put it in the kit bag that was turned inside out so the riggers would know that the cute had been jumped.

We assembled on the signal panels and did a quick head count. A deuce and a half truck was waiting to take us to the airfield. We were curious about our plane. It had a set of stairs pulled up to the wing. There was a hole in the leading edge of the wing about the size of a basketball. As I looked into the hole to the inside of the wing, I saw plenty of blood, guts, and feathers. It is incredible what a large bird can do. The plane was grounded until a repair crew could work on it. Seeing the damage made us grateful to be alive and on the ground.

Snow storm extraction via Huey.

Worst case scenario: The plane could have crashed and killed everyone, or we could have jumped and only the flight crew was killed. Well, we jumped, and the plane landed safely – God gets the credit. Our God is always in control.

ISAIAH 40:31

But they who wait for the Lord shall renew their strength;

They shall mount up with wings like eagles;

They shall run and not be weary;

They shall walk and not faint.

FLINTLOCK 86 - SPECIAL OPERATIONS COMMAND EUROPE - PART ONE

We would die in a few seconds unless I acted quickly

My Green Beret team was in England at the Sculthorpe airfield, from the WWII era. We were sequestered in a home near the airfield. We had flown in during the night, and it took 12 seconds for us to exit the plane and board a bus with blacked out windows. We took a short ride and rushed off the bus and into a house. The windows were blacked out, and no one in the neighborhood knew we were there...

The previous year we had gone to Quantico to plan a mission. Again, no one knew we were there, and we only went outside for a run when the Soviet spy satellites were not overhead. For 7 days we had planned a top secret mission to infiltrate Russia. This was well received by the top brass, and our reward was to represent the United States in the Flintlock 86 War Games. So there we were....

During our isolation at Skullthorpe, only one person was allowed to come and go at our house. All of our meals were delivered, along with our requests for maps and information. Our 12 man A-team started working and memorizing each soldiers' portion of the missions.

Our first mission was to spy, gather information, and transmit it back to England from parts of Germany. Upon completion of that mission, we would revert to a D.A. (Direct Action) mission. This would be a *search and destroy* mission.

At the end of 4 days and nights we were ready to present our plans and ability to accomplish our objectives. The twelve of us had memorized each man's duty and functions in case something happened to one of us. We had maps and intel of the areas we would travel through; we even had weather forecasts.

We sent a message to the U.S. *Top Brass* committee that we were ready for a mission review. They arrived at 6:00 pm, and we were standing tall. Our gear and rucks were on display at each man's bunk. Our maps were laid out. We had a sand table with details of our direct action target. The uniforms we wore were sterilized – names and rank had been removed. For an hour we recited our mission overall, and each team members' duties. We were questioned and cross examined until they were certain that each of us knew all of our mission. They told us to proceed with this mission and to board the plane at midnight.

As soon as they left, we burned our notes in a trash can. We then cut off the borders of the maps – we had no marks on these maps, but they had our primary and alternate drop zones, and target destination. M.R.E.'s (meals ready to eat) were issued. Each one was in a water-proof vinyl package. Three would supply our minimum daily requirements. I always unpacked them, and for 1 day I would take one meat item, 3 fruits, crackers, and 3 coffees, 3 hot chocolates, 3 creams, and 3 sugars. This would be enough for me for a day of fast travel on foot, and I usually lost about 5-lbs per week. By the 3rd day I would feel close to nature.

Our ruck sacks, when packed, weighed 75 to 90 lbs., and we carried only the essentials. Our Demolition Sargent and Radio Operator divided up their explosives and radio batteries among the team members to evenly distribute the weight. I had memorized the contents and location of everything in my ruck and could find things in the dark. This kept me from unloading everything on the trail. My pants pockets held a waterproof map case and snacks to eat while moving. My webbed gear harness had ammo pouches and water. My belt held up my pants, and I carried a survival knife. One of my shirt pockets had currency in Deutsche Marks, and the other held a Gideon New Testament. I was really loaded down, and with my M-16 rifle, main parachute, reserve parachute, and ruck sack, my extra weight equaled my body weight. So I weighed 165 lbs., and my gear weighed about 165 lbs., for a total of 330 lbs. *Phew! Try walking around like that!*

We blackened our faces and looked around to make sure there was no trace of us left there so no one could determine how many people had been in this house, and nothing of what we were doing, or where we were going.

The British military bus came and took us to the tarmac at Skullthorpe where we quickly loaded into a waiting C-130 airplane that had come from Belgium. The plane was blacked out except for the dim red lights. We got in the back of the plane and started putting on our parachutes. Even as the plane started moving, we were busy. After the main chute came the reserve chute. Next, we strapped our rifles to our sides. We had already taped over the muzzle with black electrical tape so the barrel would not get clogged when we hit the ground. The tape would stay on until our first shot, which would easily penetrate it. Finally, the ruck sack would hook to our front waist and hang down towards our feet. Because we had several jump masters on our team, we were quickly safety checked and told to sit down.

This old American Sargent was our jump master that night. As we flew over the coastline of England, our plane descended to 100 feet above the water. When we approached a boat in the English Channel, the plane would fly up to 300 feet, go over the boat, and back down to 100 feet. We were flying under the detection of radar, so no one knew we were coming.

At the halfway point over the channel, the pilots advised our team leader that we had passed the *point of no return*. When we flew into France, the plane climbed and flew about 100 feet above the tree tops. The jump master gave us a yell and signal for a 20 minute warning. It was time to mentally go over everything that was about to happen. We then got news from the load master that our primary drop zone was flooded, and the plane was diverting. We were thinking, *Sure. They are just testing us.* It was no big deal as we had memorized the alternate drop zone and the assembly point on the ground. The next jump command was a 10 minute warning, and then 6 minutes. Things really started to happen next. The jump master stood where we could see him, and he yelled and motioned a series of commands: *Get ready, Stand up, Hook up, Check static line, Check equipment.* This is where it takes great alertness. We were standing front to back and were checking the equipment and the static line connections to the parachute of the man in front of us. As you turn, you are supposed to put your hand over the reserve parachute pull handle. If this chute gets popped open by mistake, and the jump door is open, it can drag you out of the plane along with anyone hooked up on the static line in front of you.

The last jumper in our stick of 12 men slapped the rear of the man in front of him, and yelled, *Okay!* This went through every jumper until the paratrooper up front yelled, *Okay!* and gave the thumbs up sign. The load master then opened the jump door and turned the procedure over to the jump master. Meanwhile, the plane had climbed to 800 feet and leveled off. The speed had decreased to 130 miles per hour.

The jump master told the lead jumper to stand by. He then gripped both sides of the door and leaned out. He looked behind to make sure no planes were following. He didn't want to send his jumper into the propeller blades of a plane behind us. He then looked forward as the load master told him one minute.

The jump master then yelled to the lead jumper, *Stand in the door.* This jumper put his hands on the outside of the door, each side. The jump master put his arm around the soldier's waist and waited for the signal lights on the drop zone to appear under the plane.

Meanwhile, the jumpers had snugged up in line just inside the jump door. The first jumper and the jump master were looking for an inverted illuminated letter 'L' on the ground. Flash lights, chem lights, or smudge pots will do.

As we got over the drop zone, the pilot turned on the green light at the door. The jump master yelled *Go!* and slapped the trooper on the rear or leg and stood back as the rest of us jumped into the night. We were supposed to put a second between each jumper, but we were faster than that. Our evaluator, who was jumping with us, didn't know we were jumping that fast, and when he looked up, we were gone. He quickly jumped and was about 200 yards behind us.

As you leave the airplane, the first thing you notice is the roar of the engines, and quickly you feel the welcome tug as your chute has reached the end of the static line and is being pulled open. You have tucked your body very tight, with elbows by your side, and feet and legs together. You are hoping that you don't spin around and end up with the parachute twisted around. Next you feel an incredible jolt in your shoulders, back, and between your legs, telling you that the chute has opened, and the propeller blast has blown it back, so you are swinging like the pendulum of a clock. You gain control by reaching for the toggles with both hands, and quickly look up in the inside of the chute to ascertain that there are no blown panels. This is the first opportunity to look around. All of this happens in about 5 seconds since you jumped from the plane. Next you look for and stay away from other jumpers. If you float into one another, you can become tangled and both crash and die.

I strained to see, as it was pitch black. We had chosen a night with no moon so that we weren't targets floating in the sky. There was no time to relax, as I then looked for the lights on the drop zone. I was looking at the horizon and hoping to see the dark out-line of the trees. That would be my signal to lower my ruck sack. When released it would hang about 15 feet below me and hit the ground first. This would take about 75 lbs less impact off of my legs, knees, and ankles. My rate of descent rapidly increased, and then I stopped. My worst nightmare had happened!

I had either floated over a lower jumper, or he had come under me. Who was to blame? I don't know. We both had toggles and could steer our chutes in a forward direction. I was standing on top of the chute of one of my Green Beret brothers. *We would die in a few seconds unless I acted quickly.* My mind quickly flashed back to pre-jump training. This scenario was something you can talk about but cannot practice. It only happens once in a thousand jumps.

My mind raced as I realized I've only got a few seconds to react or die; and not only me but the jumper below me. My air had been stolen, and I had drifted down to his parachute. Next my chute would deflate, and my weight would deflate the chute I was stand-ing on, and we would both crash.

We were about 500 feet above the ground (about the height of a 50 story building). I was supposed to run and jump off of his chute with my hand on my reserve chute handle, which I did. I looked up and my chute was reinflating, and I floated clear from my team mate. Everything had happened so fast, and I still had to focus on what's next. I saw the ground rushing at me. I had never pulled the quick release to lower my ruck sack, and now it was too late. I hit the ground like a ton of bricks. The metal brackets on the back of my ruck rode up my legs and took the skin off of my shins. I rolled with the impact, quickly stood up, and gathered up my parachute before it could reinflate and drag me across the drop zone.

I always had a calm assurance that God was with me and pro-tecting me. No matter what we did, regardless of how dangerous, I had no fear. Always, when I hit the ground after an airborne jump,

I prayed!

I then quickly put the chute in my kit bag and headed to the assembly point. My leg was hurting, and I was limping. The medic on the ground put me in the waiting ambulance and scrubbed my wounds with iodine. It hurt and probably brought tears to my eyes. I heard them talking about flying me back to England. As heavy weapons team leader on our A-team, I felt like I had to stay with my guys, and I told them that. So they bandaged my legs, gave me an aspirin, and said goodbye as I left the ambulance and joined my team.

M-24 Sniper Rifle.

Have you ever had an experience in which you could have been killed?

One of man's greatest fears is death. The unknown awaits those who do not know God, and don't belong to Him. All they can look forward to is the grave. They have no hope, no matter how *good* a life they have lived. Look to your Bible for the promises of God.

JOHN 3:16

"For God so loved the world that He gave His only begotten Son, that whosoever believeth in Him shall not perish, but have Eternal Life."

So the grave might hold your body, but you will live forever in the glory of Heaven with your Lord and Savior. By faith each of us must believe in the saving power of Christ Jesus. Just ask Him to come into your heart and ask for forgiveness of your sins. God wants you to belong to Him and is waiting to hear from you.

PSALM 23

The Lord is my Shepherd; I shall not want.

He maketh me to lie down in green pastures.

He leadeth me beside the still waters.

He restoreth my soul.

He leadeth me in the paths of righteousness for His name's sake.

Yea though I walk through the valley of the shadow of death, I will fear no evil.

For Thou art with me; thy rod and thy staff, they comfort me.

Thou prepareth a table before me in the presence of mine enemies;

Thou anointeth my head with oil; my cup runneth over.

Surely goodness and mercy shall follow me all the days of my life,

and I shall dwell in the house of the Lord forever.

The Cliff Face of Mt. Yonah

FLINTLOCK 86 - SPECIAL OPERATIONS COMMAND EUROPE - PART TWO

We hid in plain sight

In the middle of the night we had boarded a C-130 military aircraft at Skullthorpe Airfield in England. We flew into the night and passed over the White Cliffs of Dover, then across the English Channel, into France, and then over Germany. We were an American A-team of Green Berets, good enough to be chosen to represent the United States in these huge war games involving many countries.

Hurt on the jump into Germany, I was treated in an ambulance and released to proceed with my team. I was a Heavy Weapons Team Leader, and I knew that I had to go on this mission. With only a few hours before dawn, we quickly moved out! There were multiple groups of soldiers (The OPFOR, Opposing Forces), and their only job was to track us down and eliminate us from this exercise. So we put many miles behind us while it was still dark.

We entered a mountain valley with steep sides and a village below with the only road going straight through it. We climbed up one side about halfway with a view of the valley. We were concealed by small trees, and there were bushes we could hide under.

We settled in for the day, posted a guard, and dozed off. The village came to life with all of their activities, and we patiently waited. As the day came to an end, we quietly prepared for a full night's march.

After the last light in the last house went out, we waited one hour and moved out. We double timed through the village and were pleasantly surprised that no dogs barked.

Our mission was to reach a major crossroad and monitor traffic that was supposed to be coming from Russia. That night it was 45 degrees and cool as we marched single file. At midnight, we took a short break as we ran into a light rain. I quickly put on a light weight Gore-Tex jacket. It was camouflaged and purchased just for a situation such as this. As a Green Beret, you're always looking for an edge, something that puts you ahead of the game. Gore-Tex was a wonder fabric that was new to the United States and was used on the ski slopes of the world by those who could afford it. It is light weight, water proof, and breathable. It can wick away moisture and keep you warm in cold weather. It cost $148.00 at a wilderness outfitter store, and it sure earned its money that night! We moved on into the night with rain for about an hour, when a rain storm hit us. It got really cold and wet, and our Captain said to stay in place and cover up. I pulled out my poncho, covered myself and my ruck sack, and just squatted down against a tree.

David Wegmann, our demolition Sergeant, was looking for our black team tarp and got soaking wet. When we got up before daylight, David was sick! Our medic said it was hypothermia, and it was dangerous. Our radio Sergeant sent a message for a medivac, and within 45 minutes a black hawk helicopter came for him. We divided his gear and demolition equipment before he left. We assumed he was going back to England, but they took him to a base in Germany that was monitoring our air traffic. That day we hid, dried out, and prepared for another night's march.

Now there were only eleven of us, plus our evaluator, and, with better weather and a full night's march, we made about 20 miles. Before daybreak, we hid in a Christmas tree forest between two roads. All day *we hid in plain sight.* Our final destination was a patch of woods at one of the major roads. We watched and observed all traffic, day and night. Our information, radioed back to England was received "loud and clear." We were doing our job! With sight and sound discipline, our Captain and Team Sergeant had muted conversations. There were no open fires, and we made coffee with heat tablets about the size of a sugar cube.

We were advised by a radio message that Phase One of our mission was complete. Instructions were given to proceed with Phase Two, which was a Direct Action mission (seek and destroy). Our military evaluator that traveled with us to observe our activities commented: *At this phase of the exercise, if you don't cheat, you're not trying.* Well, I was prepared!

I made arrangements with a German connection in Freyburg, a small town about 30 miles away. I went there, and the evaluator went with me. A nine passenger van was available for rent, and I paid for it with team funds. My team had been contributing to a fund, and I was carrying over 1000 Deutsche Marks. We were going to cross Germany in style at a minimum of effort.

There was a large concern for us, as the Chernobyl nuclear reactor disaster had occurred recently, and clouds of radiation were floating over Europe. We were given no guidelines or precautions, and only a Radiac Meter to wear on our collars. We never even heard the results when we turned them in at the end of the mission.

Since I rented the van and arranged for the insurance, I became the driver. With a week's head start, my beard cleaned up to reveal a narrow beard, mustache, and a goatee. For this occasion, I had brought old clothes that would blend and not stand out in public.

Every day we parked the van outside the small villages, and the two of us walked in to buy supplies, food, and do intel. Our evaluator was an American soldier serving in Germany, and he spoke the language. He watched as I bought food for my men who were waiting in the woods. If spoken to, I muttered a "yah" or "nah" with a smile. People even greeted us on the streets. On the way back out of town, we always stopped at a Gasthof House. He wanted a beer, and I wanted a coffee. He teased me because my coffee cost more than his beer, but I didn't mind because I was paying.

Every evening, back at our hideout, my good friend, Sgt. Harry Moraska would lay out all the food that I had purchased. We were eating some good German food and our MRE's (Army Meals-Ready-to-eat) were not being consumed. My team mates were comrades in arms, and I started getting requests from them to purchase items in the villages. My buddy, Frank, wanted something like German cut glass. Now where could I find it, and where could Frank put it in a ruck sack without it getting broken. So here's what we did: I purchased a jar of peanut butter in a fancy cut glass jar. We ate the peanut butter, cleaned the jar, and gave it to Frank. He was happy, and we never told him that it had held peanut butter. Another team mate wanted American cigarettes. I didn't think he should smoke because cigarettes killed my Dad. So I went to the tobacco store and asked for the strongest Turkish cigarettes without filters that they had. He lit one up that night and coughed and sputtered. I told him that maybe I could do better next time, but I didn't. He wasn't happy, and I never told him.

Every night we were transmitting back to England. Our radio man and a guard would move about one click (1000 meters) from our camp site, set up, and send a message to them. The guard would help to string an antenna between two trees. Our communication was the life of the mission, so we took care of our radio man. I remember one cool night fixing him a canteen cup of hot tomato soup. He would transmit and receive an answer in Morse code. Then we would quickly break down and move out! We didn't want the opposing forces to get a fix on our location, and that's why we went out a good distance away from our team when transmitting. The news from England indicated that some teams were caught moving during the day. They really wanted to catch us and were adding more Opposing Forces against us. Because of our transportation, we were staying a day ahead of them.

Part of our training was to receive a resupply drop because we should be getting low on food. We had to be at a certain field between 11:55pm and 12:05am and wait no longer than 10 minutes. If no plane and no drop, then repeat it 24 hours later. We hid on the edge of the field until we heard the sound of a C-130 engine. There is no other sound like it. We rushed out in the open and popped our chem lights with the correct sign. The C-130 roared over us at 300 feet and pushed out a door bundle with a small parachute. We pushed our chem lights into the dirt while the small parachute was in the air. Within a few seconds it landed, and we ripped the packing apart, divided everything, and took off. It had all happened in just a few minutes, and we quickly left the area. To confuse the forces chasing us, this plane had dipped down to tree level at four or five places over a 10 mile area. This way, no one knew if or when he pushed out a door bundle/parachute.

As we moved, Harry, our *in-house chef*, said *Hey guys! This field is full of broccoli, and it's good!* It was covered with a cool layer of dew, and we were eating while moving. The next day we thought about Chernobyl and the radioactive fallout, but it was too late! With a resupply we had a lot of food, especially since I had made trips to town.

The next day we saw an older lady with a hoe over her shoulder going to work in her garden. Stopping, we gave her a case of MRE's. That's 12 meals, and she was happy. By the time she told everyone, we would be miles down the road.

No longer moving at night, we casually cruised across Germany. Going through the villages, our men were in the back of the van covered with a tarp. I could see the military presence, but they paid no attention to me. Late in the day we camped on the outskirts of a village in the woods. The next morning I walked with the evaluator and shopped for my team mates. Arriving back, we ate a good meal and rode out that afternoon.

The scenery changed, and the hills became mountains. We had reached our destination and looked down into the valley where a river flowed into a dam and power plant. Our instructions were to *Destroy or damage*. Because we had jumped from a plane, we only had enough explosives to breach the dam and flood the valley, or damage or destroy the generators and black out the region. It looked quiet and peaceful as we looked down with our binoculars. To be safe we put eyes on the target for 24 hours prior to our attack. We chose Sgt. Frank Rigelwood to stay in hiding and let us know if anything happened or changed at our target. With our team fund, we had purchased some two way radios. We gave one to Frank and left him there.

That night a storm hit, and Frank just sat there and observed. We called him late the next day, and he said a large force had moved in and surrounded our target. Frank was talking very low and said the opposing forces were all around him. We went close by at midnight, called him, and he met up with us.

We had been compromised! In their eagerness to catch us, the target destination had been leaked. Our original plan was to float explosives down the river and detonate them against the dam or into the intake system that turns the generators. That plan had to be scrubbed. So our Team Sergeant, Robert Gignilliat put on his thinking cap and came up with another plan. He dressed like an old man with an old overcoat, hat, and shoes. He had shopping bags in each hand. In the middle of the day, Sgt. Gignilliat walked down the path next to the river. He stopped every 20 or 30 feet to rest. He was humped over and shuffling back and forth. When he finally arrived at the dam, he rested on a ledge and sat for a while. When he finally got up and hobbled away, he still had two shopping bags. No one noticed that there were two bags left on the dam. The bags he now carried were empty, and he walked on through town. After a period of time with the loaded bags sitting on the dam, our evaluator walked down to the power plant. He was now wearing his Army uniform with a badge indicating his official capacity. He simply said *You guys lost!*

The soldiers that stopped him called for their lieutenant who came racing up in a jeep. Our evaluator calmly said, *Sir, go look in those bags sitting on the dam.* He turned, looked, and walked over. In the bags were demolitions, wired with a large word, *inert.* I wish I could have seen their faces. Only Sgt. Gignilliat and the evaluator were there. What joy!

Our evaluator told the Lieutenant, *You have lost – the Americans have won!*

The previous day, the evaluator and I had stopped at the Gasthof Rose Restaurant. I had my usual coffee and roamed around looking at pictures on the wall. I saw an award from the QE2 to the owner of this restaurant. It seems that he was a chef on this great ocean liner at one time. When I called him over and pointed at the award, he beamed with pride and spoke to me in English. As we were talking, I inquired if he could feed 12 men. He said *yes,* and I had money left over, so we made a deal. I would bring my team at night after the mission was over and before we went back to England. Because we would be in uniform with weapons, he put us in a back room. We all had a great meal complete with beer and wine. As other patrons in the restaurant heard about American paratroopers (they didn't know that we were Green Berets) in the back room, free drinks started flowing our way. A German carpenter talked his way back with us. His name was Rudy, and he joked, sung songs, and enjoyed our company.

The next day we returned the van and met a truck that drove us to Stuttgart. We spent the night in some barracks at the airfield, then caught a plane back to England.

It felt good to represent the United States and do good. A lot of teams failed, but we made it!

HIDING IN PLAIN SIGHT FROM GOD

You grow up going to church with your mom and dad. Dad always prayed before the supper meal. With chores to do, you receive an allowance. You were *living the good life.* After high school came college. You had freedom as never before. Stay up late, get up – no breakfast – go to class. Being cool, you had beers with your classmates on the weekends. No more church. *Living the good life.* Then came graduation, a job, marriage and children. You started going to church as a family. It became a habit. Hearing about God was not an issue in your life because you have it all. God will be there if you need Him, but it's a busy time. *Living the good life.*

Being good and living the good life is not enough. Good people still go to Hell! However, you can be saved and know it.

All have sinned, Romans 3:23

The cost of sin is death, Romans 6:23

God loves you, John 3:16

Jesus paid the price for your sin, Romans 5:8

You can receive salvation today, Romans 10:13

Golden Glacier - May 1985

NO, I'VE DONE IT ALL

It was my last day in the Army. We were *hot loading* into a Blackhawk helicopter. As a Green Beret, we were naturally airborne. That day my unit was parachuting from an Army helicopter. I was thinking of all the jumps that I had made: about 200 or more. My favorite to jump from was the C-130, the world's safest airplane. We had jumped from the doors and the tailgate ramp. Our unit was one of the first to jump from the Sky Crane helicopter. It carried a 20 foot container under and between it's legs. I remember asking the Army photographer why he was flying with us, and he replied, *If you get killed, I have to document it.* Few have jumped the *Balloon.* I did in England and earned my British Jump Wings. We jumped from a wicker basket hanging under a World War II barrage balloon tethered by a 700-ft cable.

Over the years, we jumped everything and everywhere, and mostly at night. I remember a cold night at Fort McCoy, Wisconsin when we jumped into the snow at 33 degrees below. It's bright with the moon shining on the snow. Another time, we jumped into the warm waters of the ocean. We only jumped 3 men at the time into the ocean so the waiting boats could retrieve us ahead of the sharks. One night I landed in a swamp, and no one came to get me. Another time, my team was put out of the plane at the wrong place. We landed all over a farm in Pennsylvania instead of on the drop zone. I was fooled into thinking that I was landing in water, and it turned out to be the moon shining on the young corn leaves. My leg got twisted, and I lost my hunting knife. It was 1:00 or 2:00 in the morning, and I limped towards the farm house. It was chaos when I got there. My medic had hit and rolled off of the roof. An SAS (Brit) was out by the road, hanging from a telephone pole with a bad cut on his face. That night nine of us were packed into an ambulance and headed through the back roads to a hospital.

So I was standing there with the Captain in front of that black hawk helicopter on my last day in the Army, watching my men get chuted up and thinking of all the times that I said I didn't want to be jumping after age 50. I was thinking, *Bones are brittle and muscles weary.* Six months earlier my time had been extended for six more months until retirement. The unit was going on a drug mission, and they needed a sniper/sharpshooter as an overwatch for our men. Of course, I went – they said they needed me.

The Blackhawk's blades were turning and blowing everything around us. The engine was loud, and the Captain was yelling to be heard: *Do you want to make one last jump?* I answered, **No, I've done it all.** He thought about that for a minute and then said, *You're a jumpmaster. Go up and put everyone else out.* I then went up and down 3 or 4 times until everybody had jumped. That day one of the guys broke his collarbone as the wind drove him into the side of a hill. He was 54 years old!

After 26 years, I knew I was making the right decision. I had reached the peak of my physical fitness about 5 years earlier, and I knew it was about time. It's tough to leave something you love, but life goes on. I've always said, *Life is good, and God is great!*

A CHRISTIAN GOING TO CHURCH
IS LIKE JUMPING OUT OF AN AIRPLANE

1. Prepare for the day.

   A. A prayer and a scripture verse for the day.

   B. You get your parachute and reserve chute. You check the small rigger books to assure that they are correct and up-to-date. There is a small book that is folded and tucked into each parachute. It has the name of the rigger that packed it and the date that it was packed. The chute has to be jumped or repacked every 6 months.

2. Prepare to go.

   A. Going to church – look good – you're going into the presence of God. - Be Sharp!

   B. Inspect your chutes before you put them on. Have a jump master check your chute after you put it on.

3. Get ready

   A. Pray for yourself and others. Someone is always worse off than you and needs your prayers.

   B. Keep an eye on your fellow jumper. Does his gear look ok?

Does he?

4. Go

    A. You don't deserve it, but Christ died for you. Through grace are you saved. It's a gift from God.

    B. Exit the plane in a good tight position. Pray someone packed your chute correctly and that it opens. It's out of your hands now. Just trust. Feel the tug. Check your panels and watch for other jumpers. Hit the ground with a *Thank you* to God. Day or night.

2 TIMOTHY 4:7

"I have fought the good fight.

I have finished the race.

I have kept the faith."

AND FINALLY, A LIFE LESSON FOR YOU

I'm a good person. I do good works. Will that save me?

EPHESIANS 2:8 & 9

For by grace you have been saved through faith, and that not of yourselves; it is the gift of God, not of works lest anyone should boast.

Don't take a chance!

ROMANS 3:23

For all have sinned and fallen short of the glory of God.

ROMANS 6:27

For the wages of sin is death, but the gift of God is eternal life in Christ Jesus our Lord.

Salvation is Free

ROMANS 10:13

For whoever calls on the name of the Lord shall be saved.

JOHN 3:16

For God so loved the world that He gave His only begotten Son, that whoever believes in Him should not perish, but have everlasting life.

If you do not get saved...

REVELATION 21:18

But the cowardly, unbelieving, abominable, murderers, sexually immoral, sorcerers, idolaters, and all liars shall have their part in the lake which burns with fire and brimstone, which is the second death.

Yes.... you could be in this group! You can be saved

ROMANS 10:13

For whoever calls on the name of the Lord shall be saved.

MORE LIFE LESSONS TO FOLLOW

www.ingramcontent.com/pod-product-compliance
Lightning Source LLC
Chambersburg PA
CBHW070931120626
46546CB00004B/1385